Ildenice Nogurira Monteiro
Victor Elias Mouchrek
Odair Monteiro

Cinnamomun zeylanicum essential oil in the control of R. microplus

Ildenice Nogurira Monteiro
Victor Elias Mouchrek
Odair Monteiro

Cinnamomun zeylanicum essential oil in the control of R. microplus

Chemical composition and evaluation of the carrapathic activity of C. zeylanicum essential oil in the control of R. microplus

ScienciaScripts

Imprint

Any brand names and product names mentioned in this book are subject to trademark, brand or patent protection and are trademarks or registered trademarks of their respective holders. The use of brand names, product names, common names, trade names, product descriptions etc. even without a particular marking in this work is in no way to be construed to mean that such names may be regarded as unrestricted in respect of trademark and brand protection legislation and could thus be used by anyone.

Cover image: www.ingimage.com

This book is a translation from the original published under ISBN 978-613-9-68308-6.

Publisher:
Sciencia Scripts
is a trademark of
Dodo Books Indian Ocean Ltd. and OmniScriptum S.R.L publishing group

120 High Road, East Finchley, London, N2 9ED, United Kingdom
Str. Armeneasca 28/1, office 1, Chisinau MD-2012, Republic of Moldova, Europe
Printed at: see last page
ISBN: 978-620-8-20796-0

SUMMARY

I DEDICATE THIS WORK TO

*To my mother, **Eunice dos Santos Costa Nogueira,** who has always been by my side at every moment of my life, providing me with moments of extreme fulfillment.*

*To my father, **Nestor de Jesus Nogueira**, for his love and the opportunities he gave me.*

*To my husband **Odair,** for his affection and support during not only these two years, but also for all the times I needed his understanding and words of encouragement as I worked on this project.*

*To my dear children, **Matheus and Indira,** who give me inspiration and strength to face life's difficulties.*

*To all my **brothers**, who together make life lighter.*

To

*my **brothers-in-law and nephews-in-law**, who contribute to the joy of our family.*

THANKS

*To **God**, for everything.*

*To **Prof. Dr. Odair dos Santos Monteiro**, for his invaluable help and for his always reliable co-supervision in the preparation of this work.*

*To **Prof. Dr. Victor Elias Mouchrek Filho**, for guiding this work, his help and his friendship over the years.*

*To Prof. **Dr. Livio Martins Costa Junior**, for his help in preparing this work, for his sound guidance and for his friendship.*

*To **Jéssica**, for her collaboration and uncompromising friendship.*

*To **Aldilene**, for her valuable help in preparing this work and for her friendship and patience.*

*To **Karla Maiaquias**, for her help and collaboration in carrying out this work.*

*To my colleague **Alberto Jorge**, for his discussion and collaboration in this work.*

*To my fellow Master's students, especially my friends **Gilson, Charles and Mariane**, for their support, friendship and sharing of information.*

*To my friends **Sheyia and Luciana**, for their strength and trust.*

*To **the Natural Products Engineering Laboratory LEPRON / UFPA**, especially **Prof. Dr. José Guilherme Maia and Prof. Dr.ª . Eloisa Andrade**, for their help and collaboration in carrying out this work.*

***CEGEL and CEM Dr. João Bacelar Portela**, for their support and recognition of this work.*

To all those who directly or indirectly contributed to the completion of this work, my sincere thanks.

"Give me the hope of changing the future, and you'll drive me crazy"

Zargwill

SUMMARY

The search for natural, biologically active substances has stimulated the use of essential oils (EOs). The EO from the leaves of the species *Cinnamomum zeylanicum* Blume (cinnamon) is widely used in the pharmaceutical industry and has various biological activities, including acaricidal activity. Recently, research into the evaluation of tick activity with EOs has shown promising results in the control of the bovine tick *Rhipicephalus microplus*, reducing the use of organosynthetic products. In this context, we analyzed the chemical composition of the EO from the leaves of this plant and evaluated its tick-killing activity in the control of *R. microplus*. The chemical constituents were identified by gas chromatography coupled with mass spectrometry (GC-MS). For the carrapaticidal activity, we carried out the larvae package and engorged female immersion tests. The oil from the leaves showed a good yield and an average of 36 compounds were identified, with Benzyl Benzoate being the major component. High levels of Linalool, E-Cinnamaldehyde, α-Pinene, β-Phellandrene, Eugenol and Benzaldehyde were also observed. In the larvae test, the EO showed satisfactory larvicidal activity against *R. microplus*. In the adult immersion test, it was observed that both the EO and the benzyl benzoate standard did not cause direct mortality in engorged females, but interfered in the reproductive process of the ticks, demonstrating that the EO of this plant is a possible alternative to conventional synthetic products, since it has acaricidal action and presents partial control of the parasite.

Keywords: Cinnamon, Benzyl benzoate, Control, Bovine ticks

1 INTRODUCTION

The importance of studying natural products has been recognized over the years, and the search for these biologically active substances has stimulated the use of essential oils. Often the same active ingredient can be extracted from different plants. The choice is made bearing in mind the life cycle of the plant, the extraction yield, the nature of the compounds, the cost of the plant, the cost of maintenance and the absence of toxic compounds (FIGUEIREDO *et al.*, 2007).

In this context, essential oils represent a viable alternative in various studies involving substances of plant origin. They are related to various functions that are essential to the survival of the plant, thus playing an important role in defense against pathogenic microorganisms (OLIVEIRA *et al.*, 2006). Currently, in addition to characterization, the use of essential oils is being expanded. These oils are already used in the chemical industries of perfumery, pharmacology, pesticides, food, beverages, antiseptics and stimulants. For this reason, although still slowly, the number of studies into the chemical composition and biological properties, as well as the taxonomic, environmental and cultivation factors that lead to variation in both the quantity and quality of these essences, is growing. Brazil is considered to be a major producer and exporter of essential oils and some of their pure components. The most widely produced essences are citrus, mint, eucalyptus, geranium, citronella, vetiver, rosewood and clove (PIMENTEL, 2008).

The proper quality of raw materials must be carried out on a scientific and technical basis. Essential parameters for the quality of plant raw materials can vary depending on where the material comes from. Therefore, the geographical origin, growing conditions, stage of development, harvesting, drying and storage must be known (FIGUEIREDO *et al.*, 2007).

Research into the evaluation of insecticidal, bactericidal, fungicidal and, recently, carrapathic activities with essential oils has shown interesting results. However, the correct botanical characterization of plant species with pharmacological activity and the study of their chemical composition, with the identification, isolation and dosage of their constituents is of the utmost importance (CUNHA; RIBEIRO; ROQUE, 2007).

The use of aromatic plants to control ticks has been the focus of research in several countries. The bovine tick *Rhipicephalus microplus* is undoubtedly one of the most important parasites for Brazilian livestock (CAMPOS *et al.*, 2012).

Various research projects related to the control of *R. microplus* are being carried out, such as: the development of vaccines, pasture rotation, crossbreeding with resistant cattle breeds, biological control with some types of fungus and the use of homeopathy and plant extracts. However, most farmers only use or have access to chemical products to control this ectoparasite. Constant application of synthetic pesticides and indiscriminate disposal of the residual tick solution have been responsible for increased tick resistance to the pesticides used, poisoning of animals and applicators, tick residues in animal products and contamination of soil and water. In this context, there is an immediate need for an alternative to the organosynthetic products currently used (CAMPOS *et al.*, 2012).

In this way, the use of essential oils from aromatic plants to control *R. microplus* can alleviate the problems caused by this ectoparasite, thus reducing the use of organosynthetic products that are toxic to animals, humans and the environment. In addition, essential oils could act as sources of molecules for the synthesis of new carrapaticides, reducing the dependence of livestock farmers on organosynthetic products on the market (CAMPOS *et al.*, 2012). In this context, the aim is to study the tick-killing activity of the essential oil of the species *Cinnamomum zeylanicum* Blume.

The *cinnamon* plant (*Cinnamomum zeylanicum* Blume*)* originated in Sri Lanka (formerly Ceylon), the main producer and exporter, followed by the Seychelles, Madagascar and India (LIMA *et al.*, 2005). With its mild smell and sweet, slightly spicy taste, it is widely used in the form of bark and powder as a flavoring in cooking. Essential oil can be obtained from both the bark and the leaves by steam distillation. The essential oil from cinnamon bark is rich in cinnamic aldehyde, while that from the leaves has a different composition and is a source of eugenol. The essential oils obtained from the bark and leaves are raw materials that are widely used in the food and beverage, perfumery and pharmaceutical industries (KOKETSU *et al.*, 1997).

The most important essential oils of the *Cinnamomum* genus on the world market are those obtained from *C. verum* ("*Cinnamomum bark* oil" and "*Cinnamomum leaf* oil"), *C. cassia* ("*cassia* oil") and *C. camphora* ("*sassafras* oil" and "*ho leaf* oil"). *Cinnamomum zeylanicum* Blume (*Cinnamomum verum* J. S. Presl.), known as "cinnamon-india" and "cinnamon-ceilao" (LIMA, 2005).

Brazil regularly imports significant quantities of both bark and essential oil from different countries, given the lack of commercial cultivation of this spice in the country. Climate and soil conditions affect the cinnamon plant profoundly, so that the same species or variety grown in another country can produce a bark of very different quality to that obtained in Sri Lanka, its country of origin (RIBEIRO, 2007).

To this end, this study sought to characterize the chemical composition and some physico-chemical properties of the essential oil from the leaves of *C. zeylanicum* (cinnamon) and to evaluate its carrapathic activity against *R. microplus* (bovine tick).

2. OBJECTIVES

2.1 General objective

Characterize the chemical composition of the essential oil of *Cinnamomum zeylanicum* Blume (cinnamon) leaves and evaluate its carrapathic activity against the bovine tick *Rhipicephalus microplus.*

2.2 Specific objectives

•	Physicochemical characterization of the essential oil of *C. zeylanicum* Blume;

•	Identify and quantify the compounds present in the essential oil by gas chromatography coupled with mass spectrometry (GC-MS);

•	To evaluate the carrapathic efficiency of the essential oil and the benzyl benzoate standard against *Rhipicephalus microplus* (bovine tick);

•	Establish the 50% lethal concentration (LC_{50}) for diagnosing *R. microplus* resistance to essential oil, in larval sensitivity tests and immersion tests on engorged females.

3. THEORETICAL BASIS

3.1 General aspects of the Lauraceae family, the genus *Cinnamomum and* the species *Cinnamomum zeylanicum* Blume (cinnamon)

The Lauraceae family is widely distributed throughout the tropical and subtropical regions of the planet, comprising 49 genera and 2,500 - 3,000 species (WERFF and RICHTER, 1996). It stands out among the other families for its economic importance. Some species have been used by industries to manufacture various products, but the majority of species have their use restricted to traditional communities who have empirical knowledge of how to use these plants.

The genus *Cinnamomum* (Lauraceae) comprises around 250 species, which are characterized by being small to medium-sized shrubs and trees. The species are found in tropical forests where they grow on mountains and plains in well-drained soils (JANTAM *et al.*, 2003; JANTAM *et al.*, 2008).

Cinnamomum zeylanicum Blume (syn. *Cinnamomum verum* Presl) is an evergreen tree that grows up to 9 meters tall. The trunk is about 35 cm in diameter (figure 1). The leaves are leathery, lanceolate, ribbed at the base, shiny and smooth on the upper side and light green and finely reticulated on the underside. The flowers are yellow or greenish in color, numerous and very small, grouped in branched clusters (BALMÉ, 1978; SCHIPER, 1999). Its bark and leaves are strongly aromatic. With its mild smell and sweet, slightly spicy taste, it is widely used in the form of powdered bark as a flavoring agent in cooking and perfumery.

In popular medicine, this plant has different functions against various diseases, and the essential oil stored in secretory cells found in the leaves and stem is one of the main products responsible for its pharmacological activities (MARQUES, 2001). Cinnamon and its essential oil are used as odor and taste correctors in the preparation of some medicines (LIMA *et al.*, 2005).

Figure 1. *Cinnamomum zeylanicum* (cinnamon) tree

The aerial part of the plant is practically used as a whole. The leaves are used to extract essential oils, but the most valuable part is the bark of the branches. Cinnamon can be found on the market in powdered form and in rolled-up bark measuring around 20 to 25 cm in length (MORSBACH, 1997).

Brazil imports significant quantities of cinnamon bark and essential oil, given the lack of commercial cultivation of this spice in the country. Growing cinnamon in different environmental conditions has a profound effect on the plant, so that the same species or variety grown in another country can differ from that obtained from its country of origin, with a consequent variation in the concentrations of its main substances. In cinnamon, the chemical composition varies significantly between the different parts of the plant. In general, the bark is rich in cinnamic aldehyde and the leaf in eugenol (KOKETSU *et al.*, 1997).

The chemical composition of *C. zeylanicum* essential oil includes the following substances: cinnamic acid, benzene aldehyde, cinnamic aldehyde, coumarinic aldehyde, benzyl benzonate, cymene, cineol, elegene, eugenol, felandrene, furol, linalool, methylacetone, pinene, vanillin, among others. *C. zeylanicum* has a number of medicinal properties, such as: astringent, aphrodisiac, antiseptic, carminative, digestive, stimulant, hypertensive, sedative, tonic and vasodilator (BALMÉ, 1978; SCHIPER, 1999).

3.2 Essential oils

Essential oils, as defined by ANVISA (Brazil, 1999), are volatile products of plant origin obtained by a physical process and can be presented alone or mixed together, deterpenated or concentrated. They can be found in all plant structures, more frequently in leaves, flowers and fruit and less frequently in roots, rhizomes, wood, cortex or seeds. They occur in several genera of higher and lower plants, as well as in microorganisms, and constitute a complex mixture of substances with heterogeneous chemical structures (BRENNA *et al.*, 2003).

Essential oils come from the secondary metabolism of aromatic plants. They consist of low molecular weight substances, mainly mixtures of phenylpropanoids and terpenoids, specifically monoterpenes (c_{10}) and sesquiterpenes (c_{15}), although diterpenes (c_{20}) can also be present. In addition to these, a variety of aliphatic hydrocarbons (linear, branched, saturated or unsaturated), acids, alcohols, aldehydes, acyclic esters or lactones and nitrogen and sulphur compounds have also been identified in volatile oils (BELL and CHARLWOOD, 1980). These oxygenated derivatives are generally responsible for the characteristic aroma of an essential oil and are called terpenoids (figure 2) (SERAFINI *et al.*, 2002). These compounds can be present in different concentrations, one of which is the majority, which generally determines the biological properties (SANTURIO, 2007; BARBOSA, 2010).

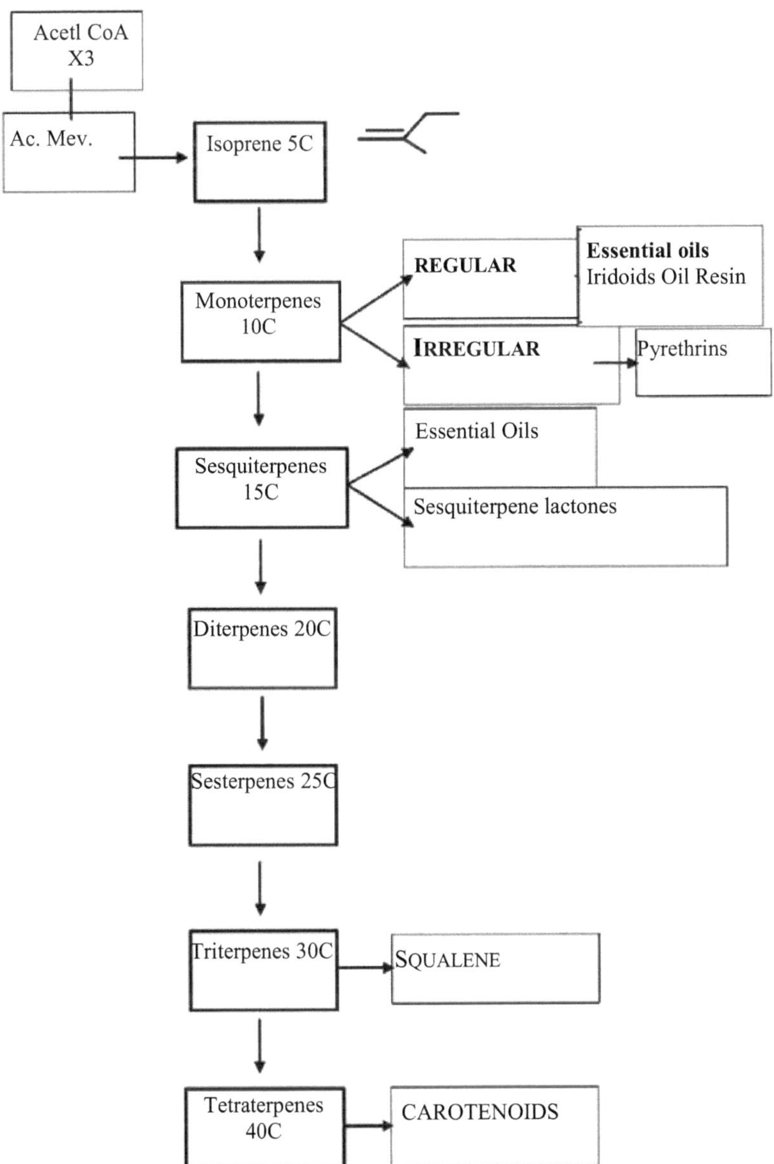

Generally, the main components determine the biological properties of the essential oil (PICHERSKY *et al.*, 2006). However, according to Mendes *et al.* (2010), the major components that are isolated in greater quantities in essential oils may reflect their biophysical and biological characteristics, but in most cases the extent of the effects of essential oils depends on the concentration of the main components modulated by other components of lesser importance, i.e. the constituents acting in synergy. For the same plant species, the number of compounds, their relative quantities and the yield of essential oils vary considerably, which can be attributed to the method of extraction of the essential oil, as well as differences in climate, location and agronomic factors such as fertilization, irrigation, harvesting and especially the stage of development of the plant at the time of harvest (BAKKALI *et al.*, 2008).

Essential oils are obtained by different processes, depending on their location in the plant, the quantity and the characteristics required for the final product. The most common methods for obtaining essential oils are: hydrodistillation, pressing or expression, blooming, extraction with organic solvents, extraction with supercritical fluid, steam distillation and microwave distillation. Among these, the most widely used method is hydrodistillation (BAKKALI *et al.*, 2008).

Volatile oils have a higher vapor pressure than water and are therefore carried away by water vapor. On a small scale, the clevenger apparatus is used for oil/water separation. After separating the volatile oil from the water, it must be dried with anhydrous sodium sulphate, for example. This procedure, although classic, can lead to thermal degradation depending on the temperature used (SANTOS, 2000).

The pressing method is widely used to obtain citrus oils and is widely used in the orange juice processing industry. Extraction methods using organic solvents can be used, but they have a number of restrictions, as they are not selective, as well as concerns about the residual part of the solvent used. Another important technological innovation is extraction using carbon dioxide in a supercritical medium. This technique is highly selective when the experimental conditions of the process are determined in advance, as well as being a totally clean technology, leaving no residue in the sample and not working at high temperatures, which reduces the risk of thermal degradation of the sample (TORRES, 2010). The great advantage of the microwave distillation method is the time taken to extract the essential oils, which is approximately 10% of the time taken for ordinary distillation.

Most commercial essential oils are analyzed by gas chromatography and mass spectrometry (BAKKALI *et al.*, 2008). Each component is identified by comparing its mass spectrum with spectra evaluated by the equipment's database, with spectra in the literature and by comparing the calculated retention indices with those already in the literature (ADAMS, 2007).

As far as quality control is concerned, essential oils have constant quality problems, which may be due to the variability of their chemical composition, adulteration or falsification, or even incorrect identification of the product and its origin (SIMÔES, 2007).

Corazza (2002) reports that essential oils generally have a lower density than water, a high refractive index and sensitivity to light and air. With regard to color, they can vary from totally colorless to strongly golden, passing through greenish, amber or yellowish nuances.

Essential oils are currently used in the pharmaceutical industry, both for human and veterinary use. They are also used as cosmetics and household products, as well as in aromatherapy. It is estimated that around 3,000 essential oils are known, of which approximately 300 are commercially important, mainly for the fragrance market (BURT, 2004).

Among the main pharmacological activities already attributed to essential oils are: antimicrobial, anti-inflammatory, antioxidant, anticholinesterase, anti-helminthic, antiparasitic, analgesic, sedative, antitumor, among others. They are also currently being used by the pharmaceutical industry as drug permeation promoters for transdermal administration (YUNES and FILHO, 2009).

3.3 Essential oils from the genus *Cinnamomum*

The composition of cinnamon essential oils can vary enormously. Previous work on *Cinnamomum zeylanicum* oil has indicated a great diversity of chemical composition, with at least five chemotypes reported: eugenol (THOMAS *et al.*, 1987; SENANAYAKE, 1978), (E)-cinnamaldehyde (VARIVAR and BANDYOPADHYAY, 1989; SENANAYAKE, 1978; BERNARD *et al.*, (1989); MOLLENBECK *et al.*, (1997), methyl benzoate (RAO *et al.*, 1988), linaloi (JIROVETZ *et al.*, (2001) and camphor (SENANAYAKE, 1978).

Nath *et al.* (1996) reported a variety of *Cinnamomum verum* grown in the Brahmaputra Valley, India, containing benzyl benzoate as the main constituent in the leaf oil (65.4%) and bark oil (84.7%), while the eugenol and cinnamic aldehyde content did not reach 1%. The presence of benzyl benzoate was first reported in cinnamon by the Wijesekera group (1974) and a probable mechanism for its formation in the plant was later proposed by Wijesekera (1978).

Morsbach *et al.* (1997) worked with essential oils from the bark and leaves of Ceylon cinnamon (*Cinnamomum verum* Presl, syn. *Cinnamomum zeylanicum* Blume). The bark and leaves came from 12 trees that had been fertilized only with organic matter (OM) or combined with chemical fertilizer (C). The average yield of essential oil was 0.2% in the bark and 2.0% in the leaves. The cinnamic aldehyde content of the bark essential oils was 54.7% (MO) and 58.4% (C). The essential oils from the leaves contained 94.1% (5 trees - MO) and 95.1% (5 trees - C) eugenol. However, the composition of the essential oils from the leaves of two different trees, one from each type of treatment, was different from most of the trees studied, showing 58.7% (MO) and 55.1% (C) eugenol, with a high safrole content (29.6% and 39.5%, respectively). No differences were observed in the composition or content of the components depending on the type of fertilization.

In the essential oils of *C. zeylanicum* from three specimens collected in the municipality of Belèm, state of Parà, (E)-cinnamyl acetate, eugenol and hydrocinnamyl acetate predominated in the leaves; benzyl benzoate and (E)-cinnamaldehyde predominated in the twig oils. In the flower oil, (E)-cinnamyl acetate predominated (MAIA *et al.*, 2007). The oil obtained in the municipality of Paço do Lumiar, in the state of Maranhao, had eugenol as its main component (87.37%) (DIAS, 2009).

The oil of *C. zeylanicum oil* showed antibacterial activities against *Escherichia coli, Staphylococcus aureus, Serratia odorifera* (DIAS, 2009), antifungal against *Candida albicans* (CASTRO, 2010), acaricidal against *Tyrophagus putrescentiae* and *Suidasia pontifica* (ASSIS, 2010) and molluscicidal against *Biomphalaria glabrata* (REIS, 2012), while its ethanolic extract has shown a natural antioxidant effect and can be used in the food, pharmaceutical and cosmetics industries (DIAS, 2009).

3.3 *Rhipicephalus microplus*

The bovine tick *Rhipicephalus microplus* (figure 3) is the only species of the *Boophilus subgenus* that occurs in Brazil. They are ticks from the Ixodidae family, known as the hard tick family. They belong to the order Parasitiformes of the class Arachnida and the suborder Metastigmata or Ixodides. The life cycle of these ticks consists of: egg, larva, nymph and adult (FLETCHMAN, 1990).

Figura 3. Image of a male *Rhipicephalus microplus* tick (A) and an engorged female still in the host's fur (B) (SEQUEIRA and AMARANTE, 2002).

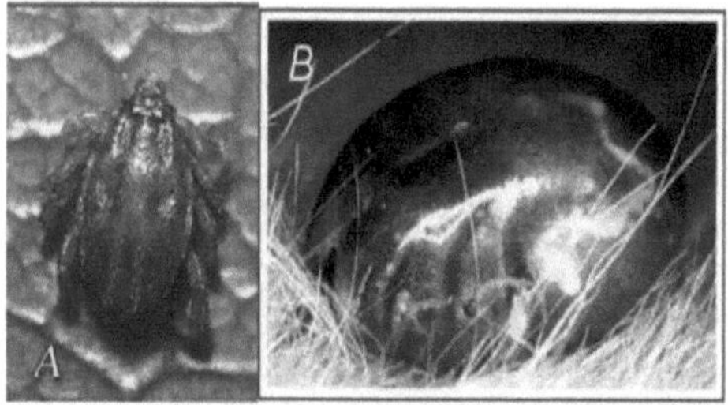

R. microplus is thought to have been introduced into most tropical and subtropical countries through imports of Asian cattle. In the Neotropics, the species is currently distributed from northern Argentina to Mexico, including the Caribbean islands and the Antilles (PEREIRA and LABRUNA, 2008).

The bovine tick is monoxenic, i.e. it needs only one host to complete its life cycle, which can be divided into a parasitic phase and a non-parasitic phase (FURLONG, 2005). The parasitic phase begins with the attachment of the larvae to the susceptible host and ends when the adults, including the fertilized and engorged females, fall off the host. The non-parasitic phase begins with the teleogyny (engorged female) after it detaches from the host and falls to the ground to oviposit. This phase ends when the larvae hatch from the eggs and access the susceptible host (PEREIRA, 1982;

PEREIRA *et al.*, 2008), as shown in figure 4.

Figura 4. Simplified diagram of the life cycle of the *Rhipicephalus microplus* tick.

Parasitic phase: (1) infective larva attaching itself to cattle; (2) nymph; (3) teleogyn in final stage of engorgement. **Free-living stage:** (4) teleogyne soon after shedding, laying in the soil; (5) eggs in the soil, incubating; (6) larva in the soil (ANDREOTTI, 2002).

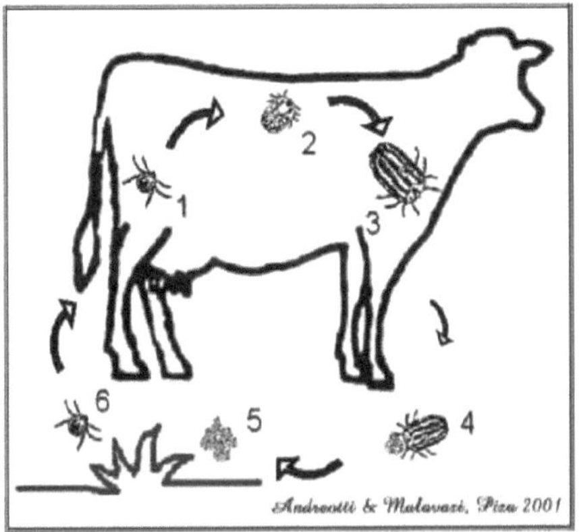

The female can lay 2,000 to 4,000 eggs. The newly hatched larvae need three to four days to form their mouthparts and then seek out and attach themselves to the hides of the host animals. They are usually very lively and climb quickly up the leaves and stems of the pasture, waiting for a host to pass by so that they can attach themselves, preferably where the skin is thinner, such as in the armpits, between the legs, genital area and under the tail. The cycle almost always begins and ends in the pasture, where the ectoparasite, the host and the environment usually interact (PEREIRA, 1982; PEREIRA *et al.*, 2008).

The cattle tick, *R. microplus*, causes economic damage to Brazilian livestock, leading to losses in milk and meat production and damage to the hide caused by inflammatory reactions at the tick's attachment sites and the transmission of diseases such as bovine parasitic sadness (caused by protozoa of the *Babesia* genus and bacteria of the *Anaplasma* genus (GRISI *et al.*; 2002).

Parasitism by the *R. microplus* tick in tropical and subtropical countries is associated with major drops in herd productivity. Conventional control of this ectoparasite has proved ineffective as a long-term control strategy, and there are recurring reports of tick populations resistant to commercially available formulations (BIEGELMEYER *et al.*, 2012). Among cattle ectoparasites, *R. microplus* infestation continues to be one of the main causes of economic losses in Brazilian livestock farming (GRAF *et al.*, 2004; RECK JÚNIOR *et al.*, 2009).

3.5 Carrapathic activity of natural products and the use of essential oils to control *Rhipicephalus microplus*

The development of chemical acaricides goes back a long way, to around 1949, when arsenical acaricides were launched. These were replaced by phosphorus compounds, amidines and, more recently, pyrethroids and chemotherapeutic insecticides (FURLONG, 2005). The indiscriminate use of these acaricides leads to problems such as the development of resistance by ticks to chemical acaricides, a fact that encourages the search for new products for this purpose.

The parasiticide market in Brazil is worth around US$ 960 million a year and accounts for 34% of the veterinary products market (SINDAN, 2010). In order to be registered, new tick control products must be at least 95% effective, according to Ministry of Agriculture and Supply (MAPA) criteria.

In Brazil, tick resistance to acaricides is widespread (OLIVEIRA *et al.*, 2000; MENDES *et al.*, 2001; MENDES, 2005). There have been reports of resistance to various compounds in the country's most important meat and milk producing states, making the situation alarming (GRAF *et al.*, 2004). Cases of resistance to carrapaticides have already been reported in several Brazilian states such as the Federal District, Goiás, Minas Gerais and Rio Grande do Sul (SILVA *et al.*, 2000; FARIAS, 1999; FURLONG, 1999; MOLENTO and DIAS, 2000; MARTINS and FURLONG, 2001).

The need to reduce the use of synthetic acaricides, since they are no longer as effective, as well as the need to create new products that control the tick population and are less harmful to the environment, makes natural products a very valuable alternative. This statement is common sense, mainly due to the fact that there are a large number of plants and their derivatives with pharmacological actions, including the acaricidal effects desired for tick control. Added to this is the risk of intoxication and increased predisposition to other diseases when cattle are bathed with chemical tickicides. In order to carry out this procedure, the animals suffer great physical stress, causing problems in meat production (CLARK, 1982).

The use of medicinal and aromatic plants to control ticks has been the focus of research in several countries. The essential oils of these plants have therapeutic, bactericidal, fungicidal and insecticidal activity and, recently, their tick activity is also being tested. Sometimes the carrapathic effect is attributed to the constituents isolated in greater quantities in the essential oil, which are the major components. However, it is possible that the activity of the main component is modulated by other compounds which are in smaller quantities (CAMPOS *et al.*, 2012).

The use of bio-pesticides derived from the secondary metabolism of plants has numerous advantages over the use of organosynthetic products. These include: they are obtained from renewable resources, they are rapidly degradable, the development of resistance to these substances made up of a combination of several active ingredients is a slow process, they do not leave residues in food and they are easily accessible and obtainable (ROEL, 2001).

Essential oils belonging to various plant species have been extensively tested to evaluate their properties as a valuable natural resource in tick control. Silva *et al.* (2009) tested the toxicity of the species Piper *aduncum* (Piperaceae) from the Amazon rainforest on engorged females and larvae of *R. microplus* and showed that the mortality of the larvae of this species was due to a constituent of the leaf oil, derived from a phenylpropanoid. For Cardona *et al.* (2007) the use of essential oil from the leaves of *Sapindus saponaria* (Sapindaceae) is a promising tool for tick control in cattle, as it causes mortality in engorged females and reduces their reproductive efficiency. Broglio-Micheletti *et al.* (2009) observed that the oil extracted from *Annona muricata* seeds was effective at low concentrations *in* controlling cattle ticks *in vitro*.

According to Bueno (2005), the use of natural products should be encouraged, but first they must undergo all the same safety procedures as synthetic products.

4. EXPERIMENTAL PART

This research was carried out at the Technology Pavilion of the Federal University of Maranhao (UFMA) in partnership with the Animal Parasitology Laboratory of the Center for Agricultural and Environmental Sciences (CCAA) of the Federal University of Maranhao (UFMA), Campos IV Chapadinha - MA and the Natural Products Engineering Laboratory (LEPRON / UFPA).

4.1 Analytical Equipment and Accessories

The methodology adopted involved the usual activities in an analytical treatment of aromatic plants, as well as *in vitro* tests of carrapathic activity using essential oils.

- **Industrial blender**

A METVISA industrial blender, model LAR.4, was used to grind the sample.

- **Infra-Red Moisture Analyzer**

A GEHAKA infrared moisture analyzer, model IV2500, was used to determine the plant's moisture content in order to calculate its yield.

- **Refractometer**

An AABE refractometer, model 2 WAJ, was used to measure the refractive index.

- **Thermostatic Bath**

An ultrathermostatic bath with circulator, model Q214S, QUIMIS, was used to cool the condensers in the distillation of essential oil by external circulation.

- **Clevenger extractor**

Glass Clevenger extractors coupled to 1000 mL round-bottomed flasks were used to extract the essential oil, coupled to blankets which were used as a heat source.

- **Centrifuge**

A FANEM model 206 centrifuge was used to separate the hydrolate contained in the essential oil.

- **Pycnometer**

A 1 mL pycnometer was used to determine the density of the essential oil.

- **Greenhouse**

An Eletrolab oven with photoperiod, model 121FC BOD, was used to carry out the larval sensitivity tests and the immersion test on engorged females.

- **Analytical balance**

An Electronic Balance Bioprecisa analytical balance, model FA-2104N, was used to weigh the teleogynes.

- **Vacuum compressor**

A Prismatec model 131b vacuum compressor fitted with a pipette was used to count live and dead larvae.

- **Gas Chromatograph coupled to Mass Spectrometer (GC-MS)**

The essential oil was analyzed by gas chromatography coupled with a mass spectrometer at the Natural Products Engineering Laboratory (**LEPRON / UFPA**) (Table 1).

Table 1. GC-MS system used for essential oil analysis

Chromatograph system	
Instrument	FOCUS (Thermo electron)
Auto-injector	AI 3000
Mass Spectrometer	
Instrument	DSQ II
Ionization source	Electronic impact
Software	Xcalibur
Data libraries	NIST (National Institute of Standards and Technology, Gaithersbury, USA) ADAMS (Allrued Publishing Corporation, Carol Stream, IL, 804 p., 2007)

4.2 Materials and Reagents

- Anhydrous sodium sulphate P. A, Na_2SO_4 , 99% purity; MERCK.

- Ethanol, 99% pure; MERCK.

- Triton 2%, SIGMA ALDRICH and MERCK.

- Benzyl benzoate P.A, SIGMA ALDRICH.

- DIGIPET and EPPENDORF automatic pipettors (2-10 µL, 10-100 µL).

- Test tubes (10x100 mm).

- Amber glass ampoules.

- Filter paper size 2 x 2cm (4cm).[2]

- Petri dishes.

4.3 Experimental Methodology

The experimental procedures used to carry out this work are described below.

4.3.1 Collecting the leaves of *Cinnamomum zeylanicum*

The green leaves of this plant were collected in March 2012 in the municipality of Santa Inês, Maranhao. Their exsiccates were identified by comparison with a specimen registered in the Joao Murça Pires Herbarium of the Museu Paraense Emilio Goeldi (MPEG), in Belém - PA, under registration number MG 165477. The leaves were dried under normal ventilation and then crushed and stored in polyethylene containers.

4.3.2 Essential oil extraction and yield calculation

The essential oil was extracted from the previously dried and crushed leaves of *Cinnamomum zeylanicum*. 100g of the dried material was weighed to obtain the essential oil. The oil was extracted by hydrodistillation in a continuous process over a period of 3.0 hours, using modified Clevenger-type glass systems coupled to refrigeration systems, with a condensation water temperature of around 10 °C (figure 5). The oils obtained were centrifuged and dried with anhydrous sodium sulphate, stored in amber glass ampoules in the absence of oxygen and kept in a refrigerated environment at 5 - 10 °C.

Figure 5 - Clevenger extractor coupled to a refrigeration system

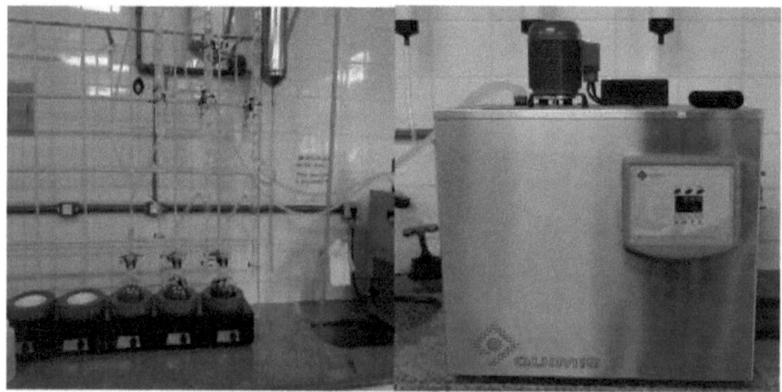

The oil yield was calculated by relating the volume of oil obtained to the mass of plant material used in the hydrodistillation on a moisture-free basis (w.b.w.), subtracting the amount of water from the

sample mass.

4.3.3 Physical characteristics of the essential oil

To characterize the physical properties of the essential oil, moisture, density, refractive index, solubility in 90% v/v ethanol, color and appearance were determined.

4.3.3.1   Humidity (%)

2.0 g of the dried plant (leaves) was weighed and placed in the infra-red moisture analyzer at an analysis temperature of 115°C for 30 minutes with a drying rate of 0.01%/min.

4.3.3.2   Density

To calculate the density, a 1.0mL pycnometer, previously dried, weighed and gauged, was used, to which the essential oil sample was added.

The following formula was used to calculate the density:

$$\text{Densidade (g/mL)} = \frac{M1 - M}{1 \text{ mL}}$$

Where: M 1 - mass of the pycnometer containing the OECZ.

M - mass of the empty pycnometer.

4.3.3.3 Solubility in ethanol at (90%)

To determine solubility, a 90% (v/v) alcohol/water mixture was used, keeping the volume of oil constant and adding increasing volumes of the alcohol mixture until it was completely solubilized.

4.3.3.4 refractive index

To determine the refractive index, glass capillary tubes were used to add the oil sample directly to the Flint prism of the refractometer, at a temperature of 25°C. The results were corrected for these experimental conditions and temperature (IAL, 2005).

4.3.3.5 Color

The technique used is visual, comparing the colors of the essences with known colors.

4.3.3.6 Appearance

The technique used is also visual, where the essences are compared in terms of their transparency.

4.4 Analysis of Essential Oil Components

For the sample, 0.1 µL of oil in hexane (10 ηg on the column) was injected into the GC-MS system under the conditions described in Table 2.

Table 2. Parameters for the GC-MS analysis of the essential oil.

Instrument	DSQ II
Column	DB-5 MS silica capillary column (30m x 0.25 mm x 0.25 μm)
Drag gases	Helium
Drag gas flow	1.2 mL/min
Injection method	Splitless (Split flow 20:1)
Ionization energy EIMS	70 ev
Column temperature Injector temperature Ion source temperature MS-interface temperature	60 to 240°C (variation of 3°C/min) 250° C 200° C 200° C

Each compound in the ion chromatograms was identified by comparing its mass spectrum (molecular mass and fragmentation pattern) with the spectra in the system library (NIST) and in the literature (ADAMS, 2007). The retention indices (RI) were determined using equation 4.1, which relates the retention time of a series of homologous hydrocarbons. A calibration curve was plotted with a series of n-alkanes (C -C_{824}) injected under the same chromatographic conditions as the samples.

AI(x)=100 Pz+100.[(RT(X)- RT (PZ)] / [(RT (Pz+1)-RT(Pz))] equation 4.1

AI: arithmetic retention index

Pz: arithmetic retention index of the hydrocarbon before X RT(x): retention time of the unknown compound RT(Pz): retention time of the hydrocarbon before X RT(Pz +1): retention time of the hydrocarbon after X

4.5 *In vitro* tests for tick activity

4.5.1 Artificial infestation *of Rhipicephalus microplus* in calves

Three four-month-old Holstein-Zebu crossbred calves were used to maintain the *Rhipicephalus microplus* population. These procedures were approved by the ethics committee for animal use and experimentation, under CEUA/UFMA registration no. 23115018061/2011-01 (figure 6). Each calf was infested every two weeks with 4000 tick larvae and the engorged females were collected 21 days after infestation. During the infestations, the calves were kept in suspended stalls with water and mineral salt, and fed elephant grass and balanced feed. These trials were carried out at the Center for Agricultural and Environmental Sciences of the Federal University of Maranhao, Campus IV

Chapadinha - MA.

Figure 6 Artificial infestation *of Rhipicephalus microplus* in calves

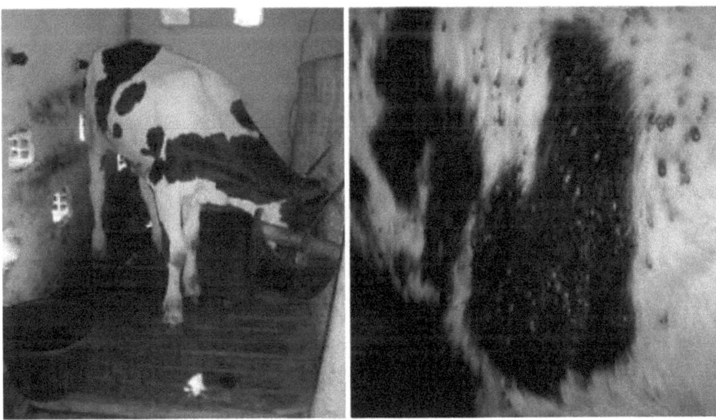

4.5.2 Larval sensitivity test

The larval sensitivity test was carried out according to the technique developed by Stone and Haydock (1962) and adapted by FAO (1984) and Leite (1988). Approximately 100 tick larvae between 14 and 21 days old were used. These were placed between two filter papers measuring 2 x 2cm (4cm^2), impregnated with 400 µl of each concentration of essential oil and benzyl benzoate standard (Sigma-Aldrich), which formed a "sandwich". This sandwich was placed inside a 7.5 x 7.5 cm non-impregnated filter paper envelope and sealed with plastic nails, according to the methodology. The envelopes were placed in a BOD oven with a temperature of 27 ± 1 °C and RH ≥ 80%, for a period of 24 hours. After this period, the live and dead larvae were counted using a vacuum compressor fitted with a pipette. Four replicates were used for each treatment and the controls were the diluent used to prepare the essential oil (triton 2%). Concentrations of 50, 25, 10, 5 and 1 mg.mL^{-1} of the essential oil and the standard benzyl benzoate were tested. The results of the larval sensitivity test were corrected according to Abbott's formula (Abbott, 1925).

The lethal concentrations (LC$_{50}$) and the confidence interval were calculated for the essential oil using the GraphPad Prism 5.0 software.

4.5.3 Immersion test for engorged females

The immersion test on *Rhipicephalus microplus* females was carried out according to the technique developed by Drummond *et al.* (1973). The teleogines (sensitive strain) collected from the artificially infested calves were washed in running water, dried on paper towels, and weighed in groups of 10 specimens, trying to get the weights of the groups as homogeneous as possible.

Each group of ticks was submerged for 5 minutes in 5 mL of the different concentrations of essential oil and benzyl benzoate standard (Sigma-Aldrich), 75, 50, 25, 10 and 5 mg.mL^{-1} . The control was the diluent used to prepare the essential oil (triton 2%). After this period, the engorged females were dried on paper towels, placed in Petri dishes and taken to a BOD oven with a temperature of 27 ± 1 °C and RH ≥ 80%, for 23 days, when the eggs were collected, weighed and placed in disposable syringes modified for this use. The syringes were also placed in a BOD oven at 27 ± 1 °C and RH ≥ 80% for a further 25 days to allow the larvae to hatch. After hatching, the larvae were visually assessed to obtain the hatching percentage. Data such as the weight of the teleogynous larvae (PT), the weight of the eggs (PO) and the hatching percentage (%E) were evaluated and the percentage of product efficiency (PE) was then calculated (Drummond *et al.*, 1973).

$$ER = \frac{PO}{PT} \times \%E \times 20000*$$

* = Number of larvae per gram of eggs.

$$EP = \frac{ER\ Control\ Group - ER\ Treated\ Group}{Control\ Group\ RE} \times 100$$

5. RESULTS AND DISCUSSION

5.1 Yield and physical characteristics of essential oil

The essential oil obtained from the leaves of *C. zeylanicum* by hydrodistillation showed a yield of 1.03% in relation to the weight of dry material used (b.l.u.). The yield value for the oil and some characteristics (density, refractive index, solubility, color and appearance) are shown in Table 3, and compared with data from the literature.

Table 3. Physical characteristics of essential oil extracted from *Cinnamomum zeylanicum* leaves.

Properties Physico-chemical	Essential oil[(a)]	Essential oil ()[b]	Essential oil [c]
Yield (%)	1,03	1,3	1,1
Density (g. mL)$^{-1}$	1,055	1,023	1,048
refractive index (N $_D^{25}$)	1,533	1,533	1, 533
Solubility in ethanol	1:1 90% (v/v)	1:1 70%(v/v)	1:1 70%(v/v)
Color	Yellow	Yellow	Yellow
Appearance	Clean	Clean	Clean

[(a)] Author's data; (^b)Reis (2011); (^c)Dias (2009)

Comparing the values for the essential oils of *C. zeylanicum* leaves with those in the literature, it can be seen that there is a similarity between them with regard to the parameters analyzed. Small differences in the values found can be attributed to factors such as time of collection, chemotype, different types of soil, conditions and storage time of the leaves, among others.

5.2 Chemical composition of the essential oil of *Cinnamomum zeylanicum* leaves

The chemical constituents were analyzed by gas chromatography coupled with mass spectrometry (GC-MS). Identification was based on analysis of mass spectra and retention index (RI) in comparison with literature data (ADAMS, 2007).

A total of 36 different compounds were identified with concentrations ranging from 0.04 to 65.39%, corresponding to 99.88% of the oil's total composition, which is shown in Table 4, with 41.62% aliphatic monoterpenes, 5.55% aromatic monoterpenes, 30.52% sesquiterpenes, 13.87% phenylpropanoids and 8.32% aromatic esters. With percentages above 2%, the main compounds found in *C. zeylanicum* leaf oil were Benzyl Benzoate (65.39%), Linalool (5.37%), E-Cinnamaldehyde (3.97%), α-Pinene (3.95%), β-Phellandrene (3.42%), Eugenol (3.36%) and Benzaldehyde (2.68%).

Table 4. Constituents identified in *Cinnamomum zeylanicum* essential oil.

Constituents	IR*	%
α-Pinene	932	3,95
Camphene	946	1,68
Benzaldehyde	952	2,68
β-Pinene	974	1,51
Mirceno	988	0,46
α-Felandrene	1002	0,09
α-Terpinene	1014	0,34
p-Cymene	1020	0,2
β-Felandrene	1025	3,42
γ-Terpinene	1054	0,04
Terpinolene	1086	0,08
Linalool	1095	5,37
cis-p-Menth-2-en-1-ol	1118	0,06
Camphor	1141	0,05
2-ethenyl-3-methyl phenol	1166	0,21
Ethyl benzoate	1169	0,18
Borneol	1165	0,23
Terpinen-4-ol	1174	0,19
α-Terpineol	1186	0,43
Z-Cinnamaldehyde	1217	0,05
E- Cinnamaldehyde	1267	3,97
δ-Elemene	1335	0,31
α-Cubebene	1345	0,09
Eugenol	1356	3,36
3-phenyl-propan-1-ol acetate	1359	0,14
α-Copaene	1374	1,05
β-Elemene	1389	0,18
E-Caryophyllene	1417	1,4
Cinnamyl E-Acetate	1443	1,27
α-Humulene	1452	0,31
D-Germacrene	1484	0,27
Bicyclogermacrene	1500	0,53
δ-Candinene	1522	0,1
Spatulenol	1577	0,13
Humulene epoxide	1608	0,16
Benzyl benzoate	1759	65,39
Aliphatic monoterpenes		41,62
Aromatic monoterpenes		5,55
Sesquiterpenes (hydrocarbons and oxygenates)		30,52
Phenylpropanoids		13,87
Aromatic esters		8,32
Total		**99,88**

* IR: Adams library retention index

The chromatogram of the essential oil extracted from the leaves of the *C. zeylanicum* species *is* shown in Figure 7 following the order of elution.

Figure 7. Chromatogram of the essential oil of the species *Cinnamomum zeylanicum*

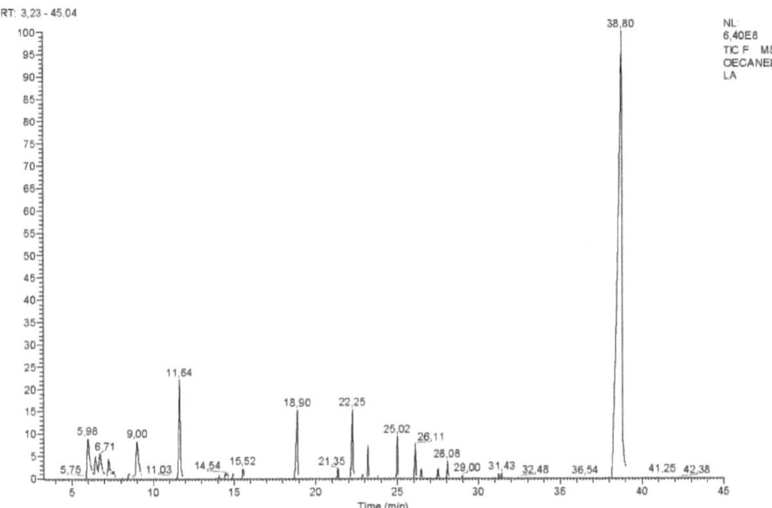

Peak 36 of the chromatogram, with a retention time of 38.80 min, corresponds to benzyl benzoate. All the peaks in the chromatogram of the essential oil of *C. zeylanicum* leaves were identified using their mass spectra.

The high content related to peak 36 (TR=38.80) of the chromatogram (65.39%), later identified as benzyl benzoate, is noteworthy, making this compound the major component of the essential oil.

Below are the mass spectra of the main peaks shown in Figure 7, following the order of the retention time, and the respective identification proposals through comparison with data from the literature (ADAMS, 2007) and from the NIST and Adams spectrographs.

Figure 8. Mass spectra of (A) compound with TR=5.98 from the chromatogram in figure 7 and (B) proposed identifications using NIST and ADAMS spectrographs (α-Pinene).

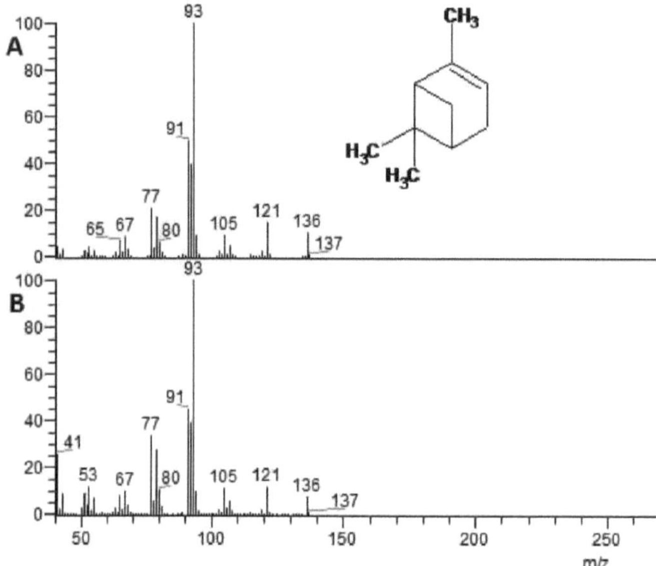

The mass spectra in figure 8 indicate, based on the literature (ADAMS, 2007) and the NIST and ADAMS spectrographs, the presence of the α-pinene compound in the essential oil. The molecular ion peak shows m/z=136, confirming the formula $C_{10}H_{16}$. The m/z=93 peak for α-pinene is probably produced by a structure with the formula $C_7H_9^+$ formed by isomerization, followed by allylic cleavage (SILVERSTEIN *et al.*, 2007).

Figure 9. Mass spectra of (A) compound with TR=11.64 from the chromatogram in figure 7 and (B) proposed identifications using NIST and ADAMS spectrographs (Linaloi).

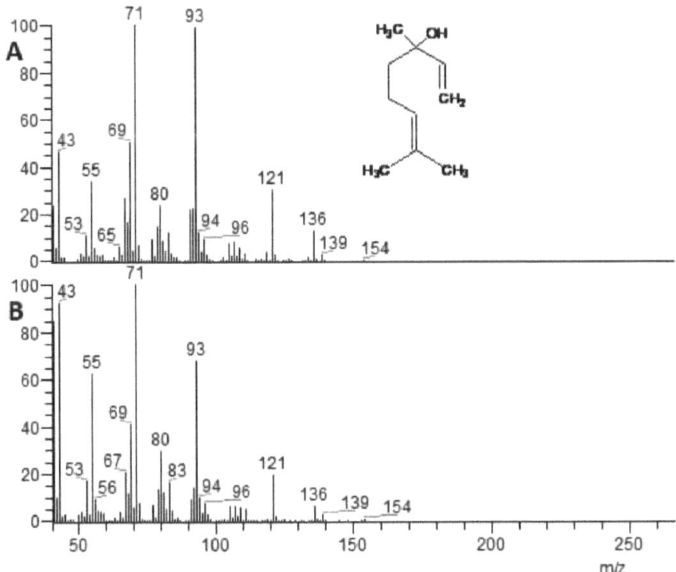

The mass spectra in figure 9 indicate, based on the literature (ADAMS, 2007) and the NIST and ADAMS spectrographs, the presence of the compound linalool in the essential oil. According to the literature, the molecular ion peak is difficult to visualize in the case of tertiary alcohols, which is the case with linalool. For this compound, whose molecular formula is $C H_{1018} O$, the molecular ion peak is m/z =154 [M]. The 136 [M - 18] peak corresponds to the loss of water, while the m/z = 121 [M - 18 - 15] peak corresponds to the loss of water and the methyl group. Linalool is a tertiary alcohol, and for compounds of this nature there is often carbon-carbon bond breaking near the oxygen atom, with the elimination of the largest group, which is evidenced by the peak at m/z=71 (H2C=CH-COH[+] -CH3) and by the peak at m/z=83 [(CH3)2C=CH- CH -CH22]. The peak with m/z=93 is due to the elimination of water (M - 18) and the $C H_{38}$ + group (M - 44) (SILVERSTEIN *et al.*, 2007).

Figure 10. Mass spectra of (A) compound with TR=18.90 from the chromatogram in figure 7 and (B) proposed identifications using NIST and ADAMS spectrographs (E-Cinnamaldehyde).

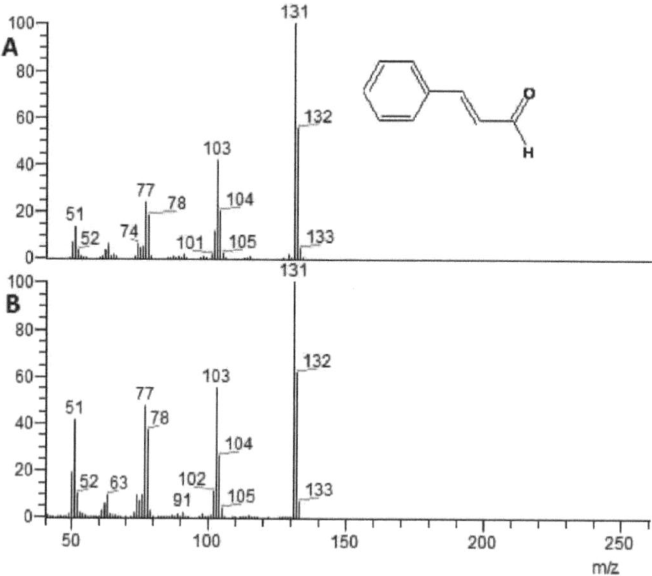

The mass spectra in figure 10 indicate, based on the literature (ADAMS, 2007) and the NIST and ADAMS spectrographs, the presence of the compound E- Cinnamaldehyde in the essential oil. The molecular ion peak shows m/z=132, confirming the formula C_9H_8O. Aromatic aldehydes exhibit intense molecular ion peaks, and the loss of a hydrogen atom by α-segmentation is a very favorable process (PAVIA *et al.*, 2010). The resulting M-1 peak can be more intense than the molecular ion peak, and this is the case for E- Cinnamaldehyde. It is also common for aromatic aldehydes to form the M-29 fragment, in the case of E-Cinnamaldehyde the peak with m/z= 103. It is possible to see in figure 10 the presence of the peak with m/z=77, corresponding to the ion $C_6H_5^+$, typical of aromatics, which in turn eliminates HC≡CH to give the ion $C_4H_3^+$ (m/z= 51) (SILVERSTEIN *et al.*, 2007).

Figure 11. Mass spectra of (A) compound with TR=38.80 from the chromatogram in figure 7 and (B) proposed identifications using NIST and ADAMS spectrographs (Benzyl Benzoate).

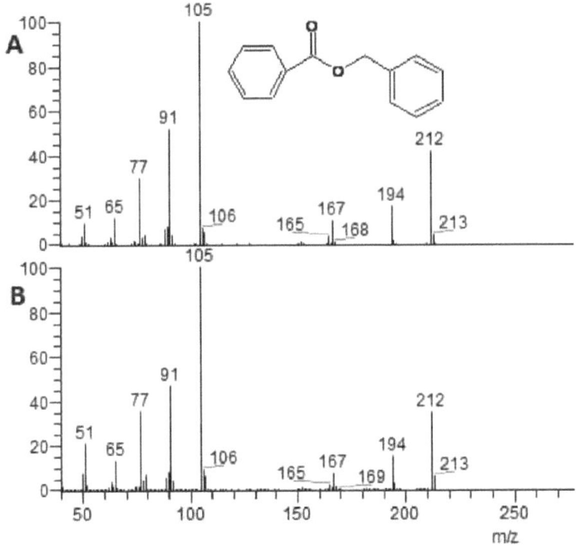

The mass spectra in figure 11 indicate, based on the literature (ADAMS, 2007) and the NIST and ADAMS spectrographs, the presence of the compound benzyl benzoate in the essential oil. The molecular ion peak shows m/z=212, confirming the formula $C_{14}H_{12}O_2$ - The peak with m/z=194 is the result of the elimination of H_2O [M - 18]. The base peak, m/z=105, is due to the formation of the stable ion $C_6H_5C\equiv O^+$, characteristic of esters. The peak with m/z=91 is characteristic of the tropylium cation ($C_7H_7^+$) and the peak with m/z=65 is the result of the neutral elimination of acetylene (C_2H_2) from the tropylium ion. The peak with m/z=77 corresponds to the ion $C_6H_5^+$, typical of aromatics (SILVERSTEIN et al., 2007).

According to previous studies on *C. zeylanicum* oil, the most common chemotype found in the leaf oil is eugenol. Other studies have shown that the oil from the leaves of *C. zeylanicum* contains significant amounts of eugenol, safrole, benzyl benzoate, (E)-cinnamyl acetate and hydrocinnamyl acetate (DIAS, 2009, LIMA et al., 2005, REIS, 2012, MORSBACH et al., 1997, NATH et al., 1996, MAIA et al., 2007). Therefore, it should be considered that these *C. zeylanicum* chemotypes may result from the plant's genotype, taking into account, above all, the season, climate and collection site.

The species of *C. zeylanicum* studied belongs to the benzyl benzoate chemotype, which, according to the literature, is used commercially as a topical sarnicidal drug that has an acaricidal action against various parasites, which suggests the potential use of this oil for this purpose (SILVA et al., 2009), so it may be of interest to traditional pharmacists. In the present study, a carrapaticidal action was obtained with this component in the control of the bovine tick *R. microplus*. In a recent study in the municipality of Olinda - PE (NEVES et al., 2009), they reported that the essential oil from the leaves of *C. zeylanicum*, containing benzyl benzoate (64.4%) as the main component, had an acaricidal action against the striped mite (*Tetranichus urticae*). *C. zeylanicum* containing a high percentage of benzyl benzoate (65.4%) in the leaf oil was collected in the Brahmaputra Valley, India (NATH *et al.*, 1996). However, although this type of leaf oil containing benzyl benzoate has been reported previously, its occurrence as a main component is little known. Glichitch, quoted by (GUENTHER, 1950) observed, based on an investigation of samples from different origins, that oils with low eugenol contents generally contained relatively large amounts of benzyl benzoate and cinnamyl esters, as is the case with the specimen studied here and also the specimens studied by (NATH *et al.*, 1996 and NEVES *et al.*, 2009), where surprisingly high values of benzyl benzoate are found for the essential oil of *C. zeylanicum* leaves.

5.3 Evaluation of larvicidal activity

In the test with *C. zeylanicum* essential oil, it was observed that at concentrations of 50, 25, 10, 5 and 1 $mg.mL^{-1}$ the larval mortality rate was 100%, 99.8%, 99%, 98.9% and 41.1%, respectively (table 5). In view of these results, the range tested was reduced to concentrations below $1mg.mL^{-1}$, where it was found that the oil was not very effective on *R. microplus* larvae. For the benzyl benzoate standard, the main component of the essential oil, larvicidal activity was tested at the following concentrations 25, 15, 10, 5, 4 $mg.mL^{-1}$, where the larval mortality rate was 100%, 100%, 100%, 85% and 68.2%, respectively (Table 5). At concentrations below 4 $mg.mL^{-1}$ it was also found that the standard did not provide satisfactory results for *R. microplus* larvae.

Analyzing the oil and the pattern of its main constituent, it was found that at concentrations starting at 5 $mg.mL^{-1}$ and 4 $mg.mL^{-1}$, a percentage of more than 50% was obtained for mortality rates, showing efficiency on *R. microplus* larvae.

Balbino *et al.* (2008) reported on the larvicidal activity of *Piper xylosteoides* essential oil, rich in safrole, α-pinene, limonene, aristolene and zingiberene, on *R. microplus* larvae, finding that at the highest concentration, this oil killed 97% of all tick larvae. In this study, α-pinene had a content of 3.97%.

The lethal concentration for 50% of the population was 1.005 $mg.mL^{-1}$ for *C. zeylanicum* oil (table 5) and 3.368 $mg.mL^{-1}$ for the standard of its major component (table 5). Comparing these mortality rates, it can be seen that the essential oil of *C. zeylanicum* leaves showed better larvicidal activity against *R. microplus*, indicating that the mixture of oil components acts more effectively than the standard alone, i.e. this suggests that the other constituents of the oil may have interacted synergistically or additively with benzyl benzoate, contributing to its toxic effects. In fact, it is often observed that the complex chemical compositions of essential oils are more effective than pure compounds (DON-PEDRO, 1996, MIRESMAILLI *et al.*, 2006, SINGH *et al.*, 2009). Combinations of compounds are generally more desirable, since not only is the spectrum of action increased, but also because various pests become more susceptible to them (SINGH *et al.*, 2009).

Apel *et al.* (2009) tested the essential oil from the leaves of five *Cunila* species at concentrations of 2.5, 5 and 10 μL.mL^{-1} ($\approx$0.25, 0.5 and 1%). *C. angustifolia* and *C. incana* caused 100% mortality of *R. microplus* larvae at the lowest concentration and *C. spicata* , at a concentration of 5 μL.mL^{-1} , *C. incisa* and *C. microcephala* had insignificant action. The main compounds found in these plants were α-pinene, β-pinene, sabinene, menthofuran and 1,8-cineole. This study also found the compounds α-pinene (3.97%), β-pinene (1.51%).

Although the mode of action of the vast majority of essential oils is not known (KIM *et al.*, 2004), the rapid action of some oils against mites is indicative of a neurotoxic action (ISMAN, 2006). The action of *C. zeylanicum* essential oil could be associated with a neurotoxic response, but further studies are needed to clarify this hypothesis.

Table 5. Percentage mortality and CL$_{50}$ of *C. zeylanicum* essential oil and Benzoate.

Benzila (standard) for *R. microplus* larvae.

Sample		Larvae			
		Mortality (%)	CL$_{50}$ (mg.mL^{-1})	95% CI	R
C. zeylanicum essential oil	Control Triton 2%	0,0			
	50 mg.mL^{-1}	100,0			
	25 mg.mL^{-1}	99,8			
	10 mg.mL^{-1}	99,0			
	5 mg.mL^{-1}	98,9	1,005	0,989-1,021	0,8421
	1 mg.mL^{-1}	41,1			
	0.95 mg.mL^{-1}	1,9			
	0.85 mg.mL^{-1}	1,6			
	0.75 mg.mL^{-1}	1,1			
Benzoate Benzila (standard)	Control Triton 2%	0,0			
	25 mg.mL^{-1}	100,0			
	15 mg.mL^{-1}	100,0			
	10 mg.mL^{-1}	100,0			
	5 mg.mL^{-1}	84,9	3,368	3,14-3,60	0,9700
	4 mg.mL^{-1}	68,1			
	3 mg.mL^{-1}	36,7			
	2 mg.mL^{-1}	1,7			

1 mg.mL^{-1}	0,0

$_{CL50}$ = Concentration (mg.mL), CI = Confidence Interval, R = Correlation Coefficient

5.4 Action on engorged females

With regard to the adult immersion test, it was observed that both the essential oil and the benzyl benzoate standard did not cause direct mortality in engorged *R. microplus* females, but they did interfere in the ticks' reproduction process, causing a decrease in egg production and a decrease in the hatchability and mortality of the larvae.

Table 6 shows that, as the concentrations of essential oil and standard benzyl benzoate increased, oviposition decreased and hatchability decreased, demonstrating that they both acted primarily on egg hatching and, consequently, their efficacy also increased, thus providing partial control over cattle ticks.

As with the larvae test, the essential oil of *C. zeylanicum* leaves proved to be more effective than the standard benzyl benzoate against *R. microplus*, a factor that can be attributed to the presence of the oil's minor components.

In the present study, as described by Gazim *et al* (2011), it was observed that the essential oils of some aromatic plants can interfere in the reproductive processes of ticks, causing a reduction in the number and weight of eggs, a reduction in the hatching of larvae and larval mortality.

No studies have been found in the literature that report the carrapathicidal activity of *C. zeylanicum* with the benzyl benzoate chemotype, but the acaricidal and insecticidal activities of the various chemical components of the essential oil from the leaves of this plant are known. A study on the relationship between monoterpene structure/activity as an acaricide against *Psoroptes cuniculi* (PERRUCI *et al.*,1995) showed a high *in vitro* activity of linalool and eugenol on this species of mite. In studies carried out by Yang *et al.* (2005), linalool and cinnamyl acetate showed insecticidal activity against *Pediculus humanus* capitis, thus confirming previous reports on the insecticidal properties of cinnamyl acetate (CHENG *et al.*, 2004). Recently (NEVES *et al.*, 2009) reported that the oil from the leaves of *C. zeylanicum,* containing benzyl benzoate, E-caryophyllene and α-copaene as its main components, exerts a fumigant action against the striped mite (*Tetranidus urticae*), which can be used in the integrated control of this mite. Oh *et al.* (2012) showed that the essential oil and extracts of *Lindera melissifolia* (Lauraceae), which has fractions of β-caryophyllene, α-humulene, germacrene-D and β-elemene, had a significant repellent effect on ticks and a moderate repellent effect on mosquitoes.

The essential oil of *Copaifera reticulata*, the copaiba tree, has carrapathic activity on *R. microplus* larvae (FERNANDES *et al.*, 2007). According to Gazim *et al.* (2011), the plant *Tetradenia riparia*

(Lamiaceae) has high carrapathic activity on *R. microplus*. These authors observed high mortality of engorged females at low concentrations of the plant's essential oil, a reduction in the number and weight of eggs, a reduction in larval hatching and larval mortality. Soares (2003) testing *Azadirachta indica* (neem) *in vitro* on engorged females of *R. (B) microplus* obtained efficacy of over 95% for both aqueous and alcoholic solutions. However, *Cymbopogon citratus* (lemongrass) was only 48% effective.

As in this experiment, there are several essential oils from various aromatic plants which, at low concentrations, are not 100% effective against *R. microplus*, but which could be used as an aid to management, since they have an acaricidal action and partially control the parasite. Therefore, the use of essential oils is an interesting option to reduce the use of chemical acaricides, although they are generally less effective than these products.

Thus, these results indicate that *C. zeylanicum* oil could represent a possible alternative for controlling *R. microplus* and encourage further studies to evaluate its efficiency on other types of mites as well.

These results suggest that the essential oil of *C. zeylanicum* is a promising alternative for the sustainable management of bovine ticks, as the use of these compounds tends to be less toxic to mammals, rapidly degrades in the environment, slowly develops resistance, and is effective and safe for applicators and consumers. However, further studies are needed to optimize the use of *C. zeylanicum* oil as an acaricidal agent.

Table 6. Efficiency *of C. zeylanicum* essential oil and Benzyl Benzoate (standard) in the test on engorged *R. microplus* females.

Sample		Engorged females			
		IPO	Red. O	Hatchability	EP (%)
	Control Triton 2%	53,2		100,0%	
	75 mg/ml	48,0	9,9%	49,6%	55,3
C. zeylanicum essential oil	50 mg/ml	48,4	9,1%	58,5%	46,8
	25 mg/ml	55,4	0,0%	69,9%	27,3
	10 mg/ml	57,6	0,0%	75,5%	18,3
	5 mg/ml	56,2	0,0%	85,0%	10,4
	Control Triton 2%	62,4		96,0%	
	25 mg/ml	53,7	14,0%	83,8%	24,9
Benzoate	15 mg/ml	57,5	7,8%	92,8%	10,9
Benzila (standard)	10 mg/ml	58,1	6,9%	94,0%	8,8
	5 mg/ml	59,0	5,4%	94,8%	6,6
	1 mg/ml	59,4	4,8%	96,3%	4,5

IPO = Egg Production Index, Red. O = Oviposition Reduction, EP = Product Efficiency

6. CONCLUSIONS

Analysis of the chemical composition of the essential oil by gas chromatography coupled with mass spectrometry (GC-MS) enabled the identification of monoterpenes, sesquiterpenes, phenylpropanoids and aromatic esters, confirming the metabolic diversity found in the *C. zeylanicum* species.

The essential oil of *C. zeylanicum* showed a composition in which Benzyl Benzoate, Linalool, E-Cinnamaldehyde, α-Pinene and β-Phellandrene were the main constituents, where Benzyl Benzoate was characterized as the chemotype.

The present study showed that the *C. zeylanicum* species provided an essential oil with a yield of 1.03% (m/m), which was considered to be high in relation to the extraction of other essential oils from aromatic plants, and the physical constants showed values similar to those obtained from the literature and the standard used for comparisons.

The larvicidal activity of the essential oil of *C. zeylanicum* leaves against *R. microplus* larvae was greater than that of the standard benzyl benzoate, a factor that can be attributed to the presence of the oil's minority components. In the adult immersion test, it was observed that both the essential oil and the benzyl benzoate standard did not cause direct mortality in engorged females of *R. microplus*, but they did interfere in the reproduction process of these ticks, causing a decrease in egg production and a decrease in the hatchability and mortality of the larvae, thus presenting partial control for the bovine tick.

Considering the results obtained, this is the first study to record the tick-killing activity of the essential oil of *C. zeylanicum* on *R. microplus*, making it a possible alternative to conventional synthetic products, which could be part of a new generation of biologically active compounds with potential for use in the sustainable control of bovine ticks. However, despite the advantages of using conventional biochar, more research is needed to prove the efficiency of these new products with these ectoparasites.

REFERENCES

ABBOTT, W.S. **A method of computing the effectiveness of an insecticide**. Journal of Economic Entomology, 1925; 18 (2): 265-267.

ADAMS, R.P. **Identification of Essential Oil Components by Gas Chromatography/Mass Spectrometry**, 4th ed. Carol Stream: Allured Publishing Corporation; 2007.

ANDREOTTI, R. **BmTI antigens induce a bovine protective immune response against *Boophilus microplus* tick**. International Immunopharmacology, 2002; vol (2): 557- 563.

APEL, M.A. **Chemical composition and toxicity of essential oils from *Cunila* species (Lamiaceae) on the** bovine **tick** *Rhipicephalus (Boophilus) microplus* . **Parasitologia Research, 2009; 105 (3): 863868,. PMID: 19421776. <u>http://dx.doi.org/10.1007/s00436-009-1455 4</u>.**

ASSIS, C. P. O. **Toxicity of essential oils on *Tyrophagus putrescentiae* (Schrank) and *Suidasia pontifica* Oudemans (acari: astigmata).** [Dissertation]. Recife: Federal Rural University of Pernambuco, 2010.

BAKKALI, F., AVERBECK, S., AVERBECK, D., IDAOMAR, M. **Biological effects of essential oils**: a review. Food Chem Toxicol, 2008; vol (46): 446 - 475.

BALBINO, J. M., GOULART, C., DIAS, J., FERRAZ, A. B. F., BORDIGNON, S., POSER, G.V., ZINI, C. A. **Chemical composition and acaricidal activity of *Piper xylosteoides* essential oil on *Rhipicephalus (Boophilus) microplus* larvae**. Salao de iniciaçao cientifica. UFRGS, Porto Alegre. 2008.

BALMÉ, F. **Plantas Medicinais**. Sao Paulo: Hemus, 1978.

BARBOSA, M. de A. ***In vitro* antimicrobial evaluation of *Punica granatum* Linn. against clinically isolated *Enterococcus faecalis***. [Monograph]. Joao Pessoa: Federal University of Paraiba, 2010.

BELL, E.A. and CHARLWOOD, B. V. **Secondary Plant Products.** New York: Spinger-Verlag, 1980.

BERNARD, T., PERINEAU, F., DELMAS, M., GASSET, A. **Extraction of essential oils by refining of plant materials. II. Processing of products in the dry state: Illicium verum Hooker (fruit) and *Cinnamomum zeylanicum* Nees (bark).** *Flav. Fragr. J.,* 1989; vol (4): 85-90.

BIEGELMEYER, P., NIZOLI, L.Q., CARDOS, F.F., DIONELLO, N.J.L. **Aspects of cattle resistance to *Riphicephalus (boophilus) microplus* ticks**. Arch Zootec, 2012; 61 (1): 11.

BRAZIL. MINISTRY OF HEALTH, Health Surveillance Secretariat. Resolution 104/99, of April

26, 1999. **Official Gazette of the Federative Republic of Brazil,** 14/05/99, 1999.

BRENNA, E., FUGANTI, C., SERRA, S. **Enantio selective perception of chiral odorants.** Tetrahedron: Asymmetry. 2003; 14 (1).

BROGLIO-MICHELETTI, S.M.F., NEVES-VALENTE, E.C., SOUZA, L.A., SILVA-DIAS, N., GIRÓN-PÉREZ, K., PRÉDES-TRINDADE, R.C. **Control de *Rhipicephalus (Boophilus) microplus* (Acari: Ixodidae) con extractos vegetales.** Rev Colombiana Entomol, 2009; vol (135): 145-149.

BUENO, O.C. **Plantas inseticidas no controle de ants cortadeiras.** Agroecologia, Botucatu, 2009; vol (28): 20-22.

BURT, S. **Essential oils: there antibacterial properties and potential applications in foods.** Int. J. Food Microbiology. 2004; vol (94): 223.

CAMPOS, R.N.S., BACCI, L., ARAÙJO, A.P.A.,, BLANK, A.F., ARRIGONI-BLANK, M.F., SANTOS, G.R.A.; RONER, M.N.B. **Essential oils from medicinal and aromatic plants in the control of the tick *Rhipicephalus microplus.*** Archivos de zootecnia. 2012; 61 (8).

CARDONA, E.Z., TORRES, F.R., ECHEVERRI, F.L. **Evaluación *in vitro* de los extractos crudos de sapindus saponaria sobre hembras ingurgitadas de *Boophilus microplus* (Acari: Ixodidae),** Scientia et Technica, 2007; vol (13): 51-55.

CASTRO, R. D. **Antifungal activity of *Cinnamomum zeylanicum* Blume (cinnamon) essential oil and its association with synthetic antifungals on *Candida* species.** [Thesis]. Joao Pessoa: Federal University of Paraiba, 2010.

CHENG, S.S., LIU, J.T., TSAI, K.H., CHEN, W.J., CHANG, S.T.,. **Chemical composition and mosquito larvicidal activity of essential oils from leaves of different *Cinnamomum osmophloeum* provenances.** J. Agric. Food Chem. 2004; 52 (14), 4395-4400.

CLARK, L.G. **Association of pesticide toxicosis with some health factors during the tick eradication program in Puerto Rico.** In: International Symposium on Veterinary Epidemiology and Economics, 1982, Arlington. Proceedings... Edwardsville: Veterinary Medicine Publishing. 1982.

CORAZZA, S. **Aromacologia: uma ciência de muitos cheiros.** Sao Paulo: SENAC, 2002.

CUNHA, A. P., RIBEIRO, J. A., ROUQUE, O. R. **"Plantas Aromâticas em Portugal - caracterizaçâo e utilizçes".** ed. Calouse Gulbenkian Foundation, Lisbon, 2007.

DIAS, V. L. N. **Phytoavailability of metals, nutritional characterization, chemical constitution, evaluation of the antioxidant and antibacterial activity of the essential oil extracted from the leaves of *Cinnamomum zeylanicum* Breyn.** [Dissertation]. Joao Pessoa: Federal University of Paraiba, 2009.

DON-PEDRO, K.N. **Investigation of single and joint fumigant insecticidal action of citrus peel oil components.** Pestic. Sci. 1996 vol (46): 79-84.

DRUMMOND, R.O., ERNST, S.E., TREVINO, J.L., GLADNEY, W.J., GRAHAM, O.H. *Boophilus annulatus* and *Boophilus microplus*. Journal of Economic Entomology, 1973; vol(66): 130-133.

FAO, **Ticks and Tickborne Disease Control**: A Practical Field Manual. FAO, Rome, 1984.

FARIAS, N.A.R. **Situación de la resistencia de la garrapata *Boophilus microplus* en la región sur de Rio Grande Del Sur, Brazil.** In: RESUMOS DO IV SEMINARIO INTERNACIONAL DE PARASITOLOGIA ANIMAL, Mérida, México. 1999. p. 25-30.

FERNADES, F.J., JORGE, I., CALVO, E.V.A., ALEJJO, E., CARBU, M., CAMAFEITA, E., GARRIDO, C., LOPEZ, J.A., JORRIN. J.; CANTORAL, J.M. **Proteomic analysis of phytopathogenic fungus *Botrytis cinerea* as a potential tool for identifying pathogenicity factors, therapeutic targets and for basic research**. Arch Microbiol, 2007; vol (187): 207215.

FIGUEIREDO, A.C., BARROSO, J.G., PEDRO, L.G. **Potencialidades e Aplicçoes Aromàticas e Medicinais**. Theoretical-Practical Course, 3ª Ed., Lisbon, Portugal, 2007.

FLETCHMANN, C.H.W. **Acaros de importância mèdico veterinària.** 3rd ed. Editora Nobel. Sao Paulo. Brazil. 1990.

FURLONG, J. **Diagnosis of the susceptibility of the cattle tick *Boophilus microplus* to acaricides in Minas Gerais state, Brazil**. In: IV International Seminar on Animal Parasitology, Puerto Vallarta, Jalisco, Mexico, 1999.

FURLONG, J. e MARTINS, J.R.S. **Resistência dos tickatos aos carrapaticidas. Embrapa Gado de Leite**. Juiz de Fora, MG. 2005.

GAZIM, Z.C., DEMARCHI, I.G., LONARDONI, M.V.C., AMORIM, A.C.L., HOVELL, A.M.C., REZENDE, C.M., FERREIRA, G.A., LIMA, E.L., COSMO, F.A.; CORTEZ, D.A.G. **Acaricidal activity of the essential oil from *Tetradenia riparia* (Lamiaceae)on the cattle tick *Rhipicephalus (Boophilus) microplus* (Acari; Ixodidae).** *Exp Parasitol*, 2011; vol (129): 175-180.

GRAF, J.F., GOGOLEWSK, N., LEACH-BING, G.A., SABATINI, M.B., MOLENTO, E.L., ARANTES, G.J. **Tick control: an industry point of view.** Parasitology, 2004; vol (129): 427-442.

GRISI, L., MASSARD, C.L., MOYA-BORJA, G. E., PEREIRA, J. B. **Impact of the main ectoparasitoses in cattle in Brazil.** A Hora Veterinària, 2002; 21 (125): 8-10.

GUENTHER, E. **Oil of *Cinnamon*. In: The Essential Oils.** New York: D. Van Nostrand, 1950; vol (4): 213-240.

ADOLFO LUTZ INSTITUTE. **Physico-Chemical Methods for Food Analysis from the Adolfo Lutz Institute.** 4 ed. Sao Paulo, 2005.

ISMAN, M.B. **Botanical insecticides, deterrents and repellents in modern agriculture and an increasingly regulated world.** Annual Review of Entomology, 2006; vol(51): 45-66.

JANTAN, I.B., YEOH, E.L.,SURIANI, R., NOORSIHA, A., ABU SAID, A. **A comparative study of the constituents of the essential oils of three *Cinnamomum* species from Malaysia.** J Essent Oil Res. 2003; vol (15): 387-391.

JANTAN, I.B., MOHARAM, B.A., SANTHANAM, J., JAMAL, J.A. **Correlation between chemical composition and antifungal activity of the essential oils of eight *Cinnamomum* species.** Pharmaceutical Biology. 2008; vol (46): 406-412.

JIROVETZ, L., BUCHBAUER, G., RUZICKA, J., SHAFI, M. P., ROSAMMA, M. K. **Analysis of *Cinnamomum zeylanicum* Blume leaf oil from south India.** *J. Essent. Oil Res.,* 13: 442-443, 2001.

KIM, H.K., KIM, J.R.; AHN, Y.J. **Acaricidal activity of cinnamaldehyde and its congeners against *Tyrophagus putrescentiae* (Acari: Acaridae).** J. Stored Prod. Res. 2001; vol (40): 55-63.

KOKETSU, M., GONÇALVES, S.L., GODOY, R.L.O. **The bark and leaf essential oils of cinnamomi (*Cinnamomum verum* Presl) grown at Paranà, Brazil.** Ciênc. Tecnol. Aliment., Campinas, 1997; 7 (3): 281-285.

LEITE, R. C. *Boophilus microplus* (Canestrini, 1887). **Susceptibility, current and retrospective use of carrapaticides in properties in the physiographic regions of Baixada do Grande Rio and Rio de Janeiro: an epidemiological approach.** [Thesis]. Rio de Janeiro: Federal Rural University of Rio de Janeiro, 1988.

LIMA, M.P., ZOGHBI, M.G.B., ANDRADE, E.H.A., SILVA, T. M.D., FERNANDES, C.S. **Volatile constituents from leaves and branches of *Cinnamomum zeylanicum* Blume (Lauraceae).** Acta Amazônica, 2005; 35 (3).

MAIA, J.G.S., ANDRADE, E.H.A., SILVA, J.K.R,; LIRA, P.N.B. **Constituintes voláteis de três espécimes de *Cinnamomum zeylanicum* Blume (Lauraceae).** 47th Brazilian Congress of Chemistry, 2007.

MARQUES, C. A.; **Economic importance of the Lauraceae family.** Floresta e Ambiente. 2001; vol (8): 195.

MARTINS, J.R. and FURLONG, J. **Avermectin resistance of the cattle tick *Boophilus microplus* in Brazil.** The Veterinary Record, 2001; 149 (92).

MENDES, M. C., VERiSSIMO, C. J., KANETO, C. N., PEREIRA, J. R. **Bioassays for measuring** the acaricides susceptibility of cattle tick *Boophilus microplus* (Canestrini,1887) in Sao Paulo State, Brazil. Arquivos do Instituto Biològico, Sao Paulo, 2001; vol (68): 23-27.

MENDES, M.C. **Resistance of the tick *Boophilus microplus* (Acari: Ixodidae) to pyrethroids and organophosphates and tick treatment on small farms**. [Thesis]. Campinas: State University of Campinas; 2005.

MENDES, S.S., BONFIM, R.R., JESUS, H.C.R., ALVES, P.B., BLANK, A.F., ESTEVAM, C.S., ANTONIOLLI, A.R, AND THOMAZI, S.M. **Evaluation of the analgesic and anti-inflammatory effects of the essential oil of *Lippia gracilis* leaves**, *J Ethnopharmacol*, 2010; vol (129): 391-397.

MIRESMAILLI, S., BRADBURY, R., ISMAN, M.B. **Comparative toxicity of *Rosmarinus officinalis* L. essential oil and blends of its major constituents against *Tetranychus urticae* Koch (Acari: Tetranychidae) on two different host plants**. Pest Manag. Sci. 2006; vol (62): 366-371.

MOLENTO, M.B. and DIAS, B. **Evaluation of the efficacy of carrapaticide products against *Boophilus microplus* in the region of Umuarama, Paranà**. Arquivos de Ciências Veterinàrias e Zoologia - UNIPAR, 2000; vol. (3): 231.

MOLLENBECK, S., KONIG, T., SCHREIER, P., SCHWAB, W., RAJAONARIVONY, J., RANARIVELO, L. **Chemical composition and analyses of enantiomers of essential oils from Madagascar.** *Flav. Fragr. J.*, 1997; vol (12): 63-69.

MORSBACH, N., KOKETSU, M., GONÇALVES, S. L., GODOY, R. L.; LOPES, D. **Essential oils from bark and leaves of *cinnamon* (*Cinnamomum verum* Presi) grown in Paranà.** *Ciências Tecnologia de Alimentos*, 1997; vol. (17).

NATH, S.C., MODON, G., BARUAH, P. **Benzyl benzoate, the major component of the leaf and stem bark oil of *Cinnamomum zeylanicum* Blume**. *J. Essent. Oil. Res.*, 1996; vol (8): 327-328.

NEVES, R. C. S., NEVES, i. A., MORAES, M. M., GOMES, C. A., BOTELHO, P. S., IÙNIOR, C. P. A.; CÂMARA C. A. G. **Atividade acaricida do óleo essencial de *Eugenia unifora* L. e *Cinnamomum zeylanicum* sobre *Tetranichus urticae*.** 32ª Reuniao Anual da Sociedade Brasileira de Quimica, Recife-PE, 2009.

NIST, Mass Spectral Library (NIST/EPA/NIH), **National Institute of Standards and Technology**, Gaithersburg, Md, USA, 2005.

OH, J., BOWLING, J.J., CARROLL, J.F. **Natural product studies of U.S. endangered plants: Volatile components of *Lindera melissifolia* (Lauraceae) repel mosquitoes and ticks.** Phytochemistry, 2012; vol (80): 28-36.

OLIVEIRA, D. R., LEITAO, G. G., SANTOS, S. S., BIZZO, H. R., ALVIANO, D. C. S., ALVIANO,

D. S., LEITAO, S. G. **Chemical and antimicrobial analyses of essential oil of** *Lippia origanoides* H.B.K. J Ethnopharmacol. 2006; vol (108).

OLIVEIRA, O.R., MENDES, M.C., JENSEN, J.R., VIEIRA-BRESSAN, M.C.R. **Determination of minimum immersion times of** *Boophilus microplus* **(Canestrini, 1887) engorged females for** *in vitro* **resistance tests with amitraz at 50% effective concentration (EC 50).** Revista Brasileira de Parasitologia Veterinària, 2000; 9 (1): 41-43.

PAVIA, D. L., LAMPMAN, G. M., KRIZ, G. S., VYVYAN, J. R. **Introduction to Spectroscopy.** 4 ed. Sao Paulo, CENGAGE. 2010.

PEREIRA, M.C. *Boophilus microplus* - **Taxonomic and morpho-biological review**. 1ª ed. Quimio Divisao Veterinària. Sao Paulo. 1982.

PEREIRA, M.C.; LABRUNA, M.B.; SZABÓ, M.P E KLAFKE, G.M. *Rhipicephalus (Boophilus) microplus.* **Biology, Control and Resistance**. Editora Med Vet. Sao Paulo. Brazil. 2008.

PERRUCCI, S., MACCHIONI, G., CIONI, P.L., FLAMINI, G., MORELLI, I. **Structure/activity relationship of some natural monoterpenes as acaricides against Psoroptes cuniculi.** J. Nat. Prod. 1995; 8 (58): 1261-1264.

PICHERSKY, E., NOEL, J.P., DUDAREEVA, N. **Biosynthesis of plant volatiles: nature's diversity and ingenuity.** *Science*, 2006; vol (311): 808-811.

PIMENTEL, F.A., CARDOSO, M.G., ZACARONI, L.M., ANDRADE, M.A., GUIMARAES, L.G..L., SALGADO, A.P.S.P., FREIRE, J.M., MUNIZ, F.R., MORAIS, A.R., NELSON, D.L. **Influence of drying temperature on the yield and chemical composition of the essential oil of** *Tanaecium nocturnum* **(barb. Rodr.) bur. and K. Shum.** Quimica Nova. 2008; 31 (3).

RAO, Y. R., PAUL, S. C., DUTTA, P. K. **Major constituents of essential oils of** *Cinnamomum zeylanicum. Indian Perfum.*, 1988; vol (32): 86-89.

RECK JÙNIOR, J.; BERGES, M.; TERRA, R.M.S.; MARKS, F.S.; VAZ JÙNIOR, I.S.; GUIMARAES, J.A.; TERMIGNONI, C. **Systemic alterations of bovine hemostasis due to** *Rhipicephalus (Boophilus) microplus* **infestation.** *Res Vet Sci,* 2009; vol (86): 56-62.

REIS, J. B. **Analytical study, toxicity evaluation and molluscicidal activity of** *Cinnamomum Zeylanicum* **Blume (cinnamon) essential oil against the** *biomphalaria glabrata* **snail.** [Dissertation]. Sao Luis: Federal University of Maranhao, 2012.

RIBEIRO, S.S.S.; SANTOS, P.A.D.; SANTOS, G.Q.; MARINHO, R.S.; KORRES, A.M.N. **Aqueous extract of** *cinnamomum zeylanicum* **Blume on** *Escherichia coli.* Proceedings of the VIII Congress of Ecology of Brazil, Caxambu - MG, 2007.

ROEL, A.R. **Utilization of plants with insecticidal properties: a contribution to sustainable rural development.** *Interaçoes: Rev Int Desenvolv Local*, 2001; vol (1): 43-50.

SANTOS, R. I. **Basic metabolism and origin of secondary metabolites.** In: SIMÔES, C. M. **Farmacognosia da planta ao medicamento.** Porto Alegre: UFRGS, 2000. 323- 354p.

SANTURIO, J.M. **Antimicrobial activity of oregano, thyme and cinnamon essential oils against Salmonella enterica serovars of poultry origin.** Ciência Rural, 2007; 37 (3): 803-808.

SCHIPER, L.P. **Secrets and virtues of medicinal plants.** Rio de Janeiro: Reader's Digest Brasil, 1999.

SENANAYAKE, U. M., LEE, T. H., WILLS, R. B. H. **Volatile constituents of cinnamon (*Cinnamomum zeylanicum*) oils.** *J. Agric. Food Chem.*, 1978; vol (26): 822-824.

SEQUEIRA, T. and AMARANTE, A. **Parasitologia animal.** CD-ROM. Sao Paulo: Epub, 2002.

SERAFINI, L. A. **Biotecnologia: avanços na agricultura e na agroindústria.** Caxias do Sul: EDUCS, 2002.

SILVA, J. K. R., SOUSA, P. J. C., ANDRADE, E. H. A., MAIA, J. G. S., J. **Agric. Food Chem.** 2009; vol (55): 9422.

SILVA, M.C.L., SOBRINHO, R.N., LINHARES, G.F.C. *In vitro* **evaluation of the efficacy of chlorfenvinphos and cyhalothrin on *Boophilus microplus* collected from cattle in the dairy basin of the micro-region of Goiânia, Goiàs.** Ciência Animal Brasileira, 2000; vol (1): 143148.

SILVERSTEIN, R. M., WEBSTER F.X., KIEMLE D.J. **Spectrometric Identification of Organic Compounds.** 7. ed. Rio de Janeiro, Livros Técnicos e Cientificos S.A., 2007.

SIMÔES, C. M. O., SCHENKEL, E. P., GOSMANN, G., MELLO, J. C. P., MENTZ, L. A. PETROVICK, P. R. **Farmacognosia: da planta ao medicamento.** Ed. 6a, Porto Alegre, ED. UFRGS, 2007.

SINDAN. **National Union of the Animal Health Products Industry, 2010.**

Veterinary market by therapeutic class and animal species, 2009. Available at: < http://www. sindan.org.br/sd/sindan/index.html >. Accessed on: May 12, 2013.

SINGH, R., KOUL, O., RUP, P.J., JINDAL, J. **Toxicity of some essential oil constituents and their binary mixtures against *Chilo partellus* (Lepidoptera: Pyralidae).** Int. J. Trop. Insect Sci. 2009; vol (29): 93-101.

SOARES, M. C. S. C. **Comparative evaluation of the efficacy of herbal medicines and chemical carrapaticide products in the control of *Boophilus microplus* (Canestrini, 1887) by means of the**

biocarrapaticidogram. [Dissertation]. Recife: Federal Rural University of Pernambuco, 2003.

STONE, B.F. and HAYDOCK, K.P. **A method for measuring the acaricide susceptibility of the cattle *B. microplus* (Can.).** Bull. Entomol. Res. 1962; vol (53): 563-578.

THOMAS, J., GREETHA, K., SHYLARA, K. S. **Studies on leaf oil and quality of *Cinnamomum zeylanicum*.** *Indian Perfum.*, 1987; vol (1): 249-251.

TORRES, F. C. **Avaliaçâo da atividade carrapaticida das fraçôes dos óleos essenciais de citronela (*C. winterianus*), alecrim (*R. officinalis*) e aroeira (*S. molle*).** [Dissertation]. Porto Alegre: Pontifical Catholic University of Rio Grande do Sul, 2010.

VARIYAR, P. S. and BANDYOPADHYAY, C. **On some chemical aspects of *Cinnamomum zeylanicum*.** *PAFAI J.*, 1989; vol(10): 35-38.

WERFF, H.W. and RICHTER, H. G. **Toward and improved classification of Lauraceae. Annals of the Missouri Botanical Garden,** 1996; vol (8): 419 - 432.

WIJESEKERA, R.O.B. **The Chemistry and Technology of *Cinnamon*.** CRC Critical Review in Food Science and Nutrition, 1978; vol (10): 1-30.

WIJESEKERA, R.O.B., JAYEWARDENE, A.L., RAJAPAKSE, L.S. **Volatile constituents of leaf, stem and root oils of cinnamon (*C. zeylanicum*).** *J. Sci. Food Agric.*, 1974; vol (5): 1211-1220.

YANG, Y.C., LEE, H.S., LEE, S.H., CLARk, J.M., AHN, Y.J. **Ovicidal and adulticidal activities of Cinnamomum zeylanicum bark essential oil compounds and related compounds against *Pediculus humanus* capitis (Anoplura: Pediculidae).** Int. J. Parasitol. Sept., 2005; vol (23).

YUNES, R.A. and FILHO, V.C. **Quimica de produtos naturais, novos fàrmacos e a moderna farmacognosia.** 2.ed.-Itajai, 2009.

More
Books!

info@omniscriptum.com
www.omniscriptum.com
OMNIScriptum

Printed by Books on Demand GmbH, Norderstedt / Germany